KISITSISINIK OQALUTTUAQ

THE NUMBER STORY

SMALL BOOK ONE

ENGLISH - KALAALLISUT / WEST GREENLANDIC

Numbers Teach Children
Their Number Names

written and illustrated by

MISS ANNA

Early Reader Edition of *The Number Story 1*
Bronze Medal Winner, 2016 Wishing Shelf Book Award

Library of Congress Control Number: 2018902040

Names: Miss Anna, author.
Title: Number story : numbers teach children their number names / Miss Anna.
Description: Portland, OR: Lumpy Publishing, 2018.
Identifiers: ISBN 978-1-949320-23-7 | LCCN 2018902040
Summary: The pictures and rhymes present stories which introduce numbers 0-10.
Subjects: LCSH Numeration—English—Kalaallisut/West Greenlandic--Pictorial works--Juvenile literature. | BISAC JUVENILE NONFICTION /
Languages: English—Kalaallisut/West Greenlandic
Classification: LCC QA141.3 .M57 2018 | DDC 513—dc23

Publisher: Lumpy Publishing
Website: www.missannabooks.com
Email: missanna@missannabooks.com

Paperback: ISBN 978-1-949320-23-7
Printed in the U.S.A.   1 3 5 7 9 10 8 6 4 2

Kisitsisit taaguutaat
ilinniarusuppigit?

It is very easy and a lot of fun!

Ajornanngitsuararsuuvoq nuannerlunilu!

Say-along our little jingle

Oqaluttuatsinik erinarsoqatigisigut!

starting from Number One!

Ataaseq aallartissutigissavarput!

1

ONE looks like my one finger.

ATAASEQ

inuama ataatsip assigaa.

ONE!
ATAASEQ!

2

TWO   trails a tail.

MARLUK

pamioq malippaa.

A TAIL! PAMIOQ!

3

THREE has bumps.

PINGASUT

tammiffeqarpoq.

BUMPY! TAMMINNARPOQ!

4

FOUR carries a sail.

SISAMAT

tingerlaat tigummivaa.

4
A SAIL!
TINGERLAAT!

5

FIVE   is a racing track.

TALLIMAT

sukkaniuttarfik assigaa.

VROOM
VROOOM!

ARFINILLIT

siuteqqutut iluseqarpoq.

SIUTEROQ!

7

SEVEN has a sharp angle.

ARFINEQ-MARLUK

inngigissumik qiverneqarpoq.

BE CAREFUL! IT'S SHARP!
MIANERSORIT! INNGIGIPPQ!

8

EIGHT   is rollercoaster rails.

ARFINEQ-PINGASUT

rutschebanetut iluseqarpoq.

YAY!
YIPPEE!

NINE is a bubble on a stick.

QULAALUAT

puaasaavoq qisuminermut
ikkusimasoq.

PUAASAQ!

10

TEN   is an eye of a whale.

QULIT

tassaavoq arferup isaa ataaseq.

WINK!
UISORERPOQ!

HELLO! ALUU!

And
Aamma

0

ZERO   is an empty pail.

NOOR'LU

qattaavoq imaqanngitsoq.

IT'S
EMPTY!
IMAQANNGILAQ!

Thank you for playing with us today.

We had a lot of fun too!

Qujanaq ullumi pinnguaqatigigatsigut.

Uaguttaaq nuannisaqaagut!

We are your Number friends,
Zero to Ten,
Who will be here for you~
Uagut kisitsisit ikinngutigaatigut illit
Noor'lu-miit Qulit tungaanut.
Illit pillutit maaniittuaannassaagut.

Bye-bye now!
See you again soon!
Paa'aj!
Takoqqiliivippugut!

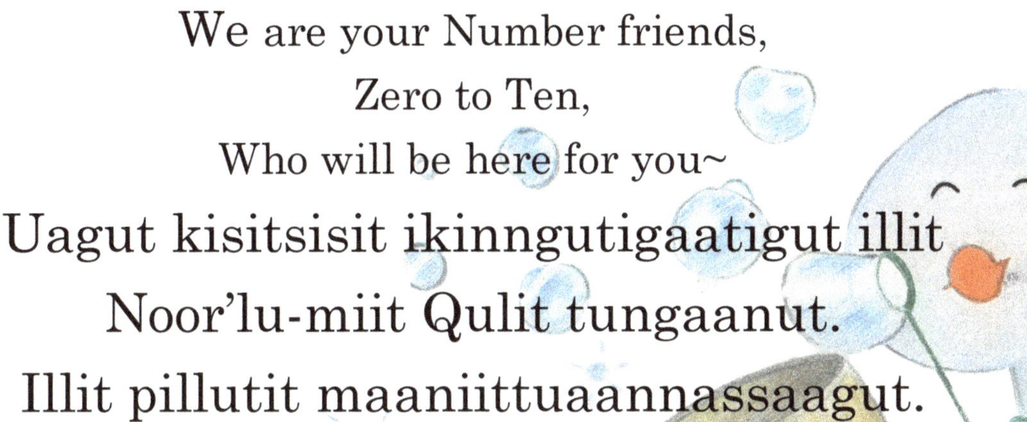

The Numbers are *SINGING* too!

To sing-a-long, look for Miss Anna Number Story
at your favorite music store like iTUNES.

MP3

| Numbers 0-10 IDENTIFYING & COUNTING | Numbers 11-20 & Ordinals first, second, third... | Numbers 0-100 & Place Values ones, tens, hundreds... | About Clocks & Telling Time hours, minutes, seconds |

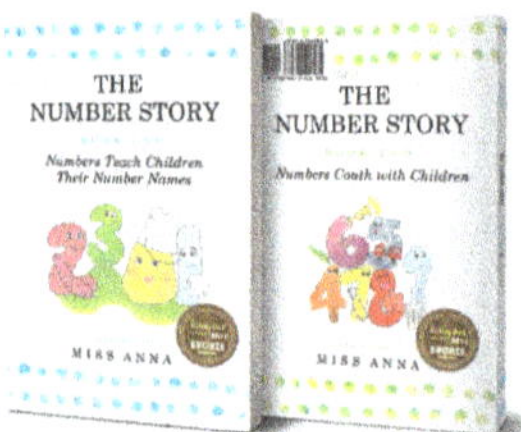   

Number Story 1 & 2
isbn: 978-0-996216-48-7

Number Story 3 & 4
isbn: 978-1-945977-01-5

Number Story 5 & 6
isbn: 978-1-945977-06-0

Number Story 7 & 8
isbn: 978-1-949320-40-4

For more Miss Anna books to love,
visit us at

www.missannabooks.com

Numbers are working hard all over the world!
Come Travel the World with Us!

www.ingramcontent.com/pod-product-compliance
Lightning Source LLC
Chambersburg PA
CBHW041100050726
47599CB00018B/2218